PROFIBUS in Practice

Installing PROFIBUS Devices and Cables

Xiu Ji

Copyright © 2013 Xiu Ji

All rights reserved.

ISBN-10: 148124521X
ISBN-13: 978-1481245210

PROFIBUS in Practice

by Xiu Ji

A textbook for the PROFIBUS International
Certified PROFIBUS Installer and Engineer courses
covering the layout, installation and testing of
PROFIBUS DP and PA networks

Xiu Ji

PROFIBUS in Practice

All text and technical details have been carefully checked. However, errors can never be completely avoided and the author accepts no liability for any damage caused by errors in this document.

Trade names and trademarks are used with no violation of the rights of their owners.

Acknowledgments

Images and illustrations used in this book are collected from many sources, e.g., the PROFIBUS Basic Slide Set (7), images used in the brochures and marketing literature of PROFIBUS International, presentations of PROFIBUS International conferences, and presentations of the PROFIBUS Group UK conferences and seminars. I thank all who have made these images and illustrations available.

I am grateful to Ann Squirrell for her help with proof reading.

Preface

PROFIBUS in Practice consists of four parts. Part I covers the installation of PROFIBUS devices and cables. Part II looks at PROFIBUS system engineering, commissioning and maintenance, Part III PROFIBUS system design and Part IV PROFIBUS operation, protocol and applications. This book is Part I of the series.

Guidelines on installing PROFIBUS devices and cables are taught as a one-day or two-day training course across the competence centres of PROFIBUS International (PI). The course is called Certified PROFIBUS Installer Course and incorporates a theory and practical test. Electricians or installers who attend the training and pass the tests are qualified as a Certified PROFIBUS Installer. The qualification is recognised world-wide and a legal requirement in many automation projects.

However, the course is intensive in terms of the content, the tutorial and examination. As a lecturer of the course for many years, I have been thinking to enhance the delivery of the course and the learning experience of participants. The result is to make use of online teaching systems. Quiz questions, course assignments and instructions are provided online and prospective participants are expected to read this book and complete quizzes and assignments online before attending face-to-face training sessions at a competence centre or at client premises. Such face-to-face sessions normally involve live demonstrations, practical activities, tutorial and assessments. I hope that the two parts (online and face-to-face) provide an enhanced and a blended approach in learning and delivering of the Certified PROFIBUS Installer course.

The book is also suitable for those who are not able to attend the face-to-face part and for technicians, engineers and students of electrical or mechanical engineering background.

Dr. Xiu Ji, July 2014

Contents

1 INTRODUCTION 1

1.1 PROFIBUS Transmission 3

1.2 PROFIBUS System Architecture 5

1.3 PROFIBUS Operation 9

1.4 PROFIBUS Applications 11

2 SEGMENTS, REFLECTIONS AND TERMINATION 13

2.1 PROFIBUS DP Segments 13

2.2 PROFIBUS PA Segments 16

2.3 PROFIBUS PA for Intrinsic Safe Applications 17

2.4 Reflections 18

2.5 Spurs 22

2.6 Causes of Reflections 25

3 CONNECTORS AND CABLING TEST 27

3.1 Connectors 27

3.2 Cabling Test 31

3.3 Address Check 32

3.4 M12 Connection 32

4 REPEATERS 34

4.1 Functions of a Repeater 34

4.2	Examples - Using Repeaters	35
4.3	ProfiHubs	37

5 ROUTING CABLES AND INSTALLING DEVICES ... 39

5.1	Interference	39
5.2	Cables for PROFIBUS DP	40
5.3	Cables for PROFIBUS PA	40
5.4	Installing PROFIBUS Cables and Devices	42

6 OPTICAL TRANSMISSION ... 50

7 SPOT THE ERRORS ... 52

8 INSTALLATION ACCEPTANCE ... 65

8.1	Network Drawing and Topology	65
8.2	Checklist for PROFIBUS DP (RS485) Grounding	67
8.3	Checklist for PROFIBUS PA (MBP) Grounding	68
8.4	Checklist for PROFIBUS DP (RS485) Cabling	69
8.5	Checklist for PROFIBUS PA (MBP) Cabling	70

9 REFERENCES ... 71

10 INDEX ... 72

11 GLOSSARY ... 73

1 INTRODUCTION

PROFIBUS is a type of fieldbus, a digital communication system for field devices. Field devices are drives, actuators, motors, pumps, valves, transmitters and sensors.

A fieldbus uses one 2-wire cable to connect many devices. A device with networking capability is often referred to as an *intelligent device*, which can contain a large quantity of data, for example, measurements, parameters and settings. In contrast to an intelligent device, a *conventional device*, for example, a temperature transmitter, transmits only one value, the measurement of the temperature.

Fieldbus is widely used in *factory automation* such as manufacturing, building and mining, and in *process automation* such as chemicals, pharmaceuticals and power plants. In factory automation, information is mainly of a discrete type, for example, on/off, open/close, and forwards/backwards, etc. Goods and objects are also discrete and thus countable, for example, cars and computers. Processing of information and logic is fast and requires little human intervention, for example, in packaging and bottling machines.

Information in process automation is mainly analogue, for example, adding 7.5 kgs of water, raising the temperature to 27.5°C and then mixing for 1.5 hours. Production in the process industry normally requires some sort of formula or a sequence of events. Information processing is slower compared to that in factory automation.

Most industries have a mix of factory and process automation (1). For example, a car assembly plant is mainly factory automation (FA). However, to assemble a car requires a lot of power so most car assembly plants also have a power generation plant, which is mainly process automation (PA).

Within the PROFIBUS family, there are two systems, namely PROFIBUS DP (Decentralised Periphery) and PROFIBUS PA (Process Automation). PROFIBUS DP is typically used in factory automation and PROFIBUS PA in the process automation. With the two variants, PROFIBUS provides a complete technology for networking field devices in many automation applications.

An automation system comprises a "controller", a number of field devices and connecting cables. The controller is in the format of a PLC (Programmable Logic Controller), DCS (Distributed Control System) or SCADA (Supervisory Control and

Data Acquisition). It is the central brain. Field devices are sensing and driving components. Between a controller and the field devices are connecting cables or "nerve systems".

Figure 1 shows an automation system with conventional wiring where multi-core cables and junction boxes are used. Communication is peer to peer which means, for example, to add a valve needs an entire wire from the valve to the controller's IO module. Safety systems, for example, emergency shutdown, safe-guarding and interlocking systems are usually separated from the normal operation systems. Thus, safety devices such as the emergency stop switches shown in Figure 1 are wired separately to a dedicated safety controller (not shown in Figure 1).

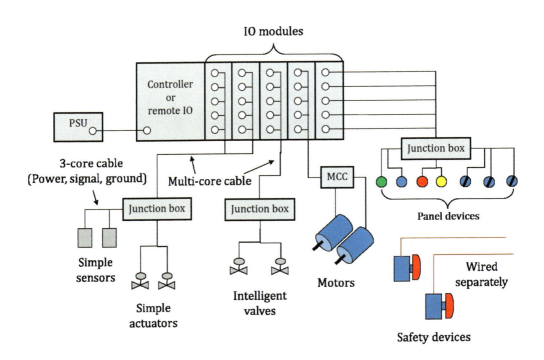

PSU: Power Supply Unit. MCC: Motor Control Centre

Figure 1 : Conventional wiring of a controller and field devices

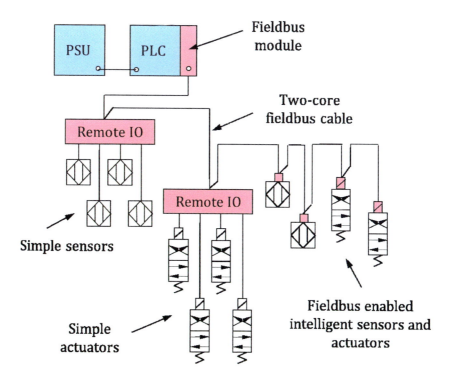

Figure 2 : Fieldbus wiring of a controller and field devices

Figure 2 shows a fieldbus wiring. One cable connects sensors and actuators (valves) to a controller. Fieldbus enabled devices are intelligent devices shown in the right-hand side of the figure and are directly connected to the bus cable. Simple sensors or actuators are connected through two remote IO stations. A *remote IO* (remote Input and Output) station, also referred to as a distributed IO station, is an intelligent device that has a PROFIBUS interface.

1.1 PROFIBUS Transmission

PROFIBUS DP and PA systems share the same communication protocol. The only difference between them is the cabling or transmission technology. PROFIBUS DP uses RS485 (Recommended Standard 485) and PA uses MBP (Manchester Bus Powered) transmission. Figure 3 shows a hierarchical view of automation systems where PROFIBUS DP and PA devices are at the bottom layer (10).

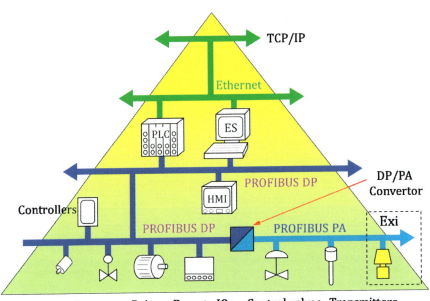

Figure 3 : PROFIBUS DP and PA devices

PROFIBUS DP and PA devices are of the same protocol and thus appear the same as nodes to a controller (DP master). However, PROFIBUS DP and PA cables are different and a convertor is required to join the two cables.

At the second layer of the hierarchy are controllers, HMI (Human Machine Interface) stations and engineering stations (ES). At the top layer are enterprise networks.

A PROFIBUS DP signal (or RS485 signal) is a square-waved voltage signal and PA signal (or MBP signal) is a current signal, as shown in Figure 4. A DP/PA convertor is required to convert between RS485 and MBP signals. There are two types of convertors, DP/PA Coupler and DP/PA Link Module.

Data is encoded as such: when a RS485 signal voltage is high the data value is 1. When the signal voltage is low the data value is 0. In MBP transmission, data is 0 when current changes from high to low; data is 1 when current changes from low to high.

A RS485 signal is between 4 and 7 volts and MBP between 1 and 19 mA as shown in Figure 4.

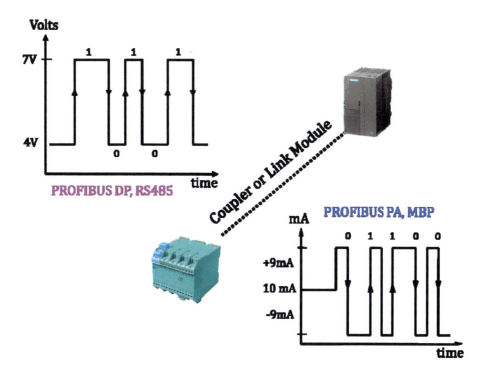

Figure 4: Conversion of PROFIBUS DP and PA signals

1.2 PROFIBUS System Architecture

A PROFIBUS system comprises at least a *master* station and several *slaves* (7). Slaves are field devices, i.e., drives, actuators, motors, pumps, valves, transmitters, and sensors. Figure 5 shows a number of DP and PA slaves.

Master is also referred to as DP Master, since the communication protocol is called PROFIBUS DP protocol. PA devices use the same protocol.

There are two types of master, Class 1 and Class 2. Controllers and some HMI stations are Class 1 masters and are permanently installed in an automation system. Class 2 masters are engineering stations or troubleshooting tools. They are not necessarily permanently installed in an automation system but added when required, for example, to change parameters and settings.

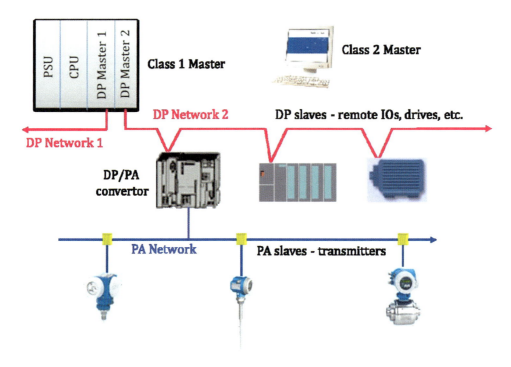

Figure 5: PROFIBUS system architecture

Class 1 masters can be in several formats, e.g., built into a controller, into a network card or into a gateway. Figure 5 shows a controller with two DP masters of Class 1. Class 2 master functions are normally of software applications and installed on a PC (Figure 5).

There are no DP masters that can connect PA slaves directly in a PA network and thus a DP/PA convertor is always required even if there are no DP slaves.

Every DP or PA device has a unique device identification number (device ID number), which is administrated by PROFIBUS International. A device ID number is issued for a type of DP or PA devices and identifies the firmware of the devices.

Slaves and Class 1 masters have a GSD (General Station Description) file, which contains device parameters, diagnostic text, and settings. A GSD file also contains the device ID number, which is expressed in a four digit hexadecimal number, e.g., 802D and D730. A GSD filename is in the format of the manufacturer's short name plus the hexadecimal ID number, e.g., SIEM802D and WAGOD730.

A GSD file is part of the device, created and supplied by device manufacturers and should be backed up and maintained by users.

In a live network, every PROFIBUS device is given an *address* through which communication is directed. The addresses must not overlap and every node must have a unique address. Station addresses can be set in three ways, which are illustrated in Figure 6.

(1) a set of local switches on the device (DIP[1] switches or rotary switches),
(2) over the PROFIBUS network using a configuration tool, and
(3) using special software and a serial link or hand-held tool.

A new address is activated only when the power to the device is reset as shown in red in Figure 6.

When there are DIP switches, Switch 1 to 7 define the device address (Figure 6, (1)). If there are more than 7 switches, the extra switches are for options other than addressing, e.g., firmware update. These options are defined in the manufacturer's manual. In the case of Figure 6, Switch 8 sets the address options; for example, Switch 8 being ON indicates software addressing and OFF indicates hardware addressing.

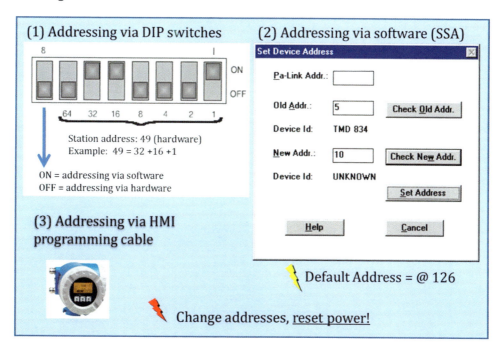

Figure 6: Addressing a PROFIBUS slave

---

[1] DIP: A DIP switch is a manual electric switch that is packaged with others in a group in a standard dual in-line package (DIP).

If software addressing is used, a PROFIBUS device must support a function called *Set Slave Address* (SSA) and has a default address at 126 (Figure 6, (2)). With an initial address at 126, an SSA tool is able to change it to a working address. In practice, a bench-top or test PROFIBUS network should be available to assign node addresses. New devices are added to the test network one by one. If two new devices are connected to the test network at the same time, both would at address 126 and the network will not work due to the duplicate addresses.

Some devices require a programming cable or point-to-point serial link to set the very first device address; this is rather specific and inconvenient (Figure 6, (3)).

There are a total of 128 addresses within a PROFIBUS network, ranging from 0 to 127, which is the capacity of one PROFIBUS network. The top address, 127, is reserved and used for broadcast. The lowest address, 0, is also reserved or purposely left un-occupied as the address is normally used for an engineering station. An engineering station is added to a network to modify parameters and settings. The address, 126, is a default address for any new device and should not be used as a working address. A network has at least one master and this master is normally at address 1. Special addresses are listed in Table 1.

PROFIBUS Address Rang: 0 – 127 (128 addresses available)	
Address 127	Reserved for broadcast and cannot be used as a working address. If a master wants to send a message to all slaves, it sends the message to 127.
Address 126	Reserved for new devices which have no DIP switches and their addresses are set over a PROFIBUS network.
Address 0 - 125	Available for masters and slaves
Address 1	Used for the main master (Class 1 Master).
Address 0	Used for an engineering station (Class 2 Master) or a diagnostic tool. Reserved even there is neither an engineering station nor a troubleshooting tool.

Table 1: Special PROFIBUS addresses

PROFIBUS DP Speed (Bit Rate)	9.6 Kbits/s 19.2 Kbits/s 45.45 Kbits/s 93.75 Kbits/s 187.5 Kbits/s 500.0 Kbits/s 1.5 Mbits/s 3.0 Mbits/s 6.0 Mbits/s 12.0 Mbits/s
PROFIBUS PA Speed (Bit Rate)	31.25 Kbits/s

Table 2: PROFIBUS network bit rates

A PROFIBUS DP network runs at one of the ten speeds as provided in Table 2. However, there is only one speed for PROFIBUS PA, which is at 31.25 Kbits/s.

1.3 PROFIBUS Operation

Slaves are configured prior to powering up of a PROFIBUS network. During power up, the master goes through a *start-up procedure* as shown in Figure 7 and then enters data exchange. If there are any errors during the "Set Parameters" or "Check Configuration" steps, the procedure will not proceed to the "Data Exchange" step but will wait for correct parameters and/or configuration to be assigned. These "pauses" are feedback loops shown in blue in Figure 7.

On being powered up, the master first sends out a diagnostic message to check which stations are on the network. In the reply from a slave, the master receives the slave's device ID number. It then sends parameters to the slave. The master then checks if the modules currently present on the slave matches what is configured in its memory through a "Check Configuration" telegram. In the second diagnostic message of a start-up procedure, the master receives information about whether the parameters and configuration are correct. If all are correct, the master and slave enter into data exchange. Data exchange is cyclic (12), which means that a master exchanges data with the slaves in its configuration one after another.

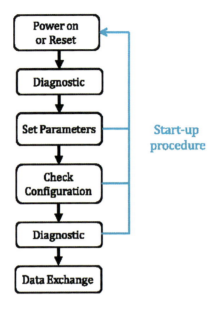

Figure 7: Start-up procedure of a PROFIBUS master or slave

When a slave is switched off and on again, the slave also goes through the start-up procedure of Figure 7. Once in data exchange, there will be neither "Set Parameters" nor "Check Configuration" step and hence parameters or configuration cannot be altered after these steps. After a start-up procedure, data exchange in PROFIBUS is cyclic, which means that the master polls slaves one by one.

In Figure 8, a master at address 1 is exchanging data with slaves at addresses 3, 4, 5 and 6. The *cycle* begins with Master 1 talking with Slave 1. Master 1 requests Slave 1 and the request telegram carries output data for Slave 1. Slave 1 responds to the request and the response telegram carries input data for Master 1. After exchanging data with Slave 1, Master 1 polls the next slave. Within a cycle the master talks to all slaves in its configuration. After one cycle the master starts a new cycle; the cycle then repeats. A Bus cycle time is determined by the number of slaves, network speed and the amount of data being exchanged.

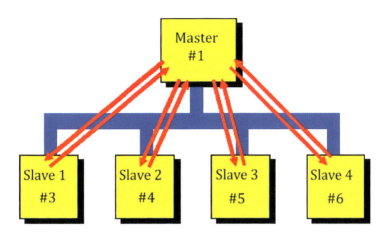

Figure 8: Cyclic communication

If there are multiple masters on a bus, there will be *token passing* between masters. When a master holds the token, it controls the network initiating requests and receiving responses. After completing its cycle, the master passes the token to the next master.

1.4 PROFIBUS Applications

At the beginning of this chapter, PROFIBUS DP is introduced as a general bus technology for factory automation and PROFIBUS PA for process automation. With standardisation at the device level through *device profile*, PROFIBUS is also widely used in functional safety systems and motion control (7).

Table 3 shows PROFIBUS solutions for various industrial applications and PROFIBUS standards, i.e., the communication protocol and device profiles.

PROFIdrive drives or motors are used in high speed applications, e.g., machine tools, robots and trajectory controls where speeds are normally synchronised.

Functional safety systems typically refer to safe guarding, interlocking and emergency stopping of automatically controlled systems.

Application	Process Automation Ex or Non-Ex Area	Factory Automation	Motion Control	Safety System
PROFIBUS Solution	PROFIBUS PA	PROFIBUS DP	PROFIBUS Drives	PROFIBUS Safety
Device Profile	PA	Not mandatory	PROFIdrive	PROFIsafe
Communication Protocol	PROFIBUS DP	PROFIBUS DP	PROFIBUS DP	PROFIBUS DP
Transmission Technology	RS 485 / RS 485 IS[2] MBP / MBP IS	RS 485	RS485	RS485

Table 3: PROFIBUS standards, solutions and applications

---

[2] IS : Intrinsic Safe or Intrinsic Safety. Refer to Section 2.3.

2 SEGMENTS, REFLECTIONS AND TERMINATION

2.1 PROFIBUS DP Segments

Stations in a PROFIBUS DP network are connected frequently in a linear structure using *sub-D connectors*. Such a connector joins two cables together and forms a continuous cable as shown in Figure 9. The two ends of the network cable are terminated, which is indicated by the symbol "T". The two terminators are to stop reflections. Reflections are discussed in Section 2.4.

As all stations are connected on one cable, it is important that a station can be replaced without disrupting the network. Thus, the sub-D connector should be adopted in PROFIBUS wiring.

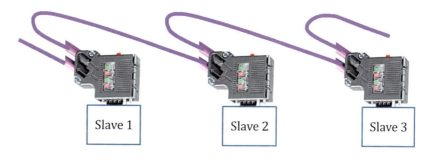

(b) Sub-D connectors forming a linear bus

(a) A linear bus with termination

Figure 9: A PROFIBUS DP network using sub-D connectors

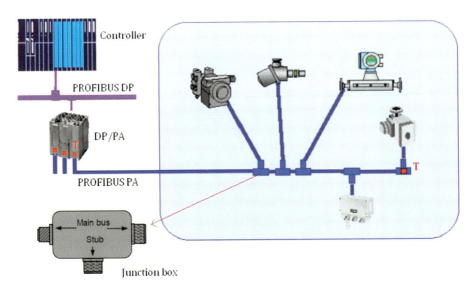

Figure 10: A PROFIBUS PA network in a tree topology

Compared to PROFIBUS DP networks, stations in PA networks are connected more flexibly, for example, using both junction boxes (Figure 10) and sub-D connectors.

The total number of addresses in a PROFIBUS network is 128, as provided in Table 1. Removing the special addresses, the working addresses for masters and slaves are from 1 to 125. Thus, a network can have up to 125 nodes. These nodes are spread over several segments since each DP or PA segment can only have up to 32 stations (Figure 11).

A PROFIBUS *segment* (DP or PA) is a continuous piece of cable which can host a maximum number of 32 stations. If there are more than 32 stations, a *repeater* is required and then two segments are formed. Over two segments, 62 PROFIBUS stations can be connected. A repeater counts as one load in a segment. The total number of stations in a fully populated segment is 32, including repeaters. However, a more realistic limit for a segment is 27 stations, which provides spares for station expansion or for connecting a temporary diagnostic tool.

Let us look at Figure 11. There are a number of slaves, a master and repeater in Segment 1. The repeater counts as one *RS485 load* and hence counts into the 32 limit. It also counts as one RS485 load in the adjacent segment, Segment 2. In Segment 2, apart from the repeater and slaves, there is also a DP/PA Coupler and an Optical Link Module (OLM) and each counts as one RS485 load.

PROFIBUS in Practice, Installing PROFIBUS Devices and Cables

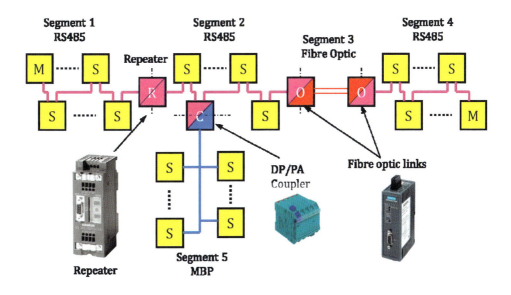

Figure 11: PROFIBUS network segments

Speed	Segment maximum cable length
9.6 Kbps	1,200 m
19.2 Kbps	
45.45 Kbps	
93.75 Kbps	
187.5 Kbps	1,000 m
500.0 Kbps	400 m
1.5 Mbps	200 m
3.0 Mbps	100 m
6.0 Mbps	
12.0 Mbps	
31.25 Kbps (PROFIBUS PA)	1,900 m (non-intrinsic safe)
	1,000 m (intrinsic safe)

Table 4: PROFIBUS segment maximum cable length

The primary function of a repeater is to re-generate and re-strengthen the RS485 signal. Therefore, when a segment load is close to 27 stations, a repeater is required.

A repeater is also required if a network cable is to cover a long distance. The maximum cable length of a DP segment is dependent on the network speed (Table 4). For example, if a network runs at 1.5 Mbps and requires 700 meters of cables, then 3 repeaters are required and 4 segments are formed, which provides the maximum cable length of 800 meters.

Apart from repeaters, Optical Link Modules (OLMs) are also used for DP network expansion. An OLM converts between the electronic and light signal, and joins a copper cable with fibres. Fibres can cover very long distances and are immune to noise. A repeater, Coupler or OLM is an RS485 load in a segment but it is not a node on the network as they do not require a network address. Repeaters that do require a network address are special repeaters such as diagnostic repeaters.

2.2 PROFIBUS PA Segments

Repeaters are not normally used in PROFIBUS PA segments since a PA segment is formed by using a DP/PA Coupler rather than a repeater. PA stations draw power from the cable and the power is supplied from a DP/PA coupler. Therefore, the number of PA stations that can be connected to a segment is determined by the power capacity of a DP/PA coupler which, in many cases, is significantly less than the limit of 32 stations. Figure 12 shows a coupler with a current capacity of 70 mA. If each of the PA slaves draws 14 mA, then the maximum number of stations which can be connected to this segment is 5. Of course, if the current rating increases to 400 mA then the maximum number of stations will rise to 28.

If there are a large number of PA slaves, these stations will be arranged into several segments and each segment is driven by a DP/PA coupler.

A DP/PA coupler is a load on the DP side and also a load on the PA side. Therefore, it counts into the 32 limit. However, a DP/PA coupler is not a node and hence does not use a network address. Some special DP/PA convertors such as the Siemens DP/PA Link Module, which is a DP slave and also a PA master do require a network address.

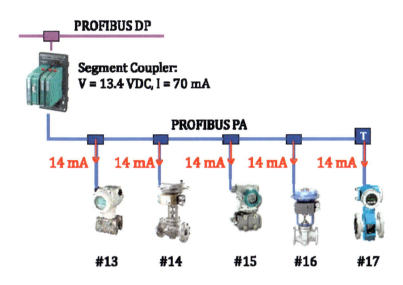

Figure 12: Power to PA stations

2.3 PROFIBUS PA for Intrinsic Safe Applications

When electrical equipment operates in a potentially explosive atmosphere (a hazardous environment) special precautions must be taken to ensure that it will not cause an explosion. Several protection methods are available, however *intrinsically safe protection*, EExi[3], has many advantages for instrumentation. Intrinsically safe protection is based upon limiting the current and voltage (and hence power) available to the field-mounted device. Capacitance and inductance are also controlled to limit the stored energy available for a spark. Different classes of protection are provided: EExib which is safe in the event of a single fault and EExia which remains safe in the event of a double fault. Special "barriers" (Figure 13) that incorporate voltage and current limiting devices are placed between the power supply and the explosive area.

---

[3] EExi: **E**uropean Standard for **Ex**plosive Applications **i**ntrinsic Protection Technique

Intrinsic safety certified devices must be used that meet the capacitance and inductance requirements for the protection category. They are placed after the barrier. Also the cable length and spur lengths must be limited. This means that when used to protect a PA segment, the number of devices and maximum cable length are significantly less than the PROFIBUS (i.e. IEC 61158-2) specification would allow for non- explosive applications.

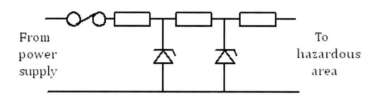

Figure 13: Typical protection barrier

PROFIBUS PA segments can be designed to be intrinsically safe by ensuring all devices are certified for intrinsic safety and that the cable and terminations meet the requirements. Traditionally the "entity concept" was used for designing intrinsically safe applications; however a relatively new method called the "Fieldbus Intrinsically Safe Concept (FISCO)", has been introduced to simplify system design. The detailed design of an intrinsically safe PA segment is beyond the scope of this book and discussed in Part III of PROFIBUS in Practice.

2.4 Reflections

Reflections occur when signals travel down a cable and encounter changes in electrical characteristics. Reflections are like echoes and can occur in any electrical transmission. However, in digital communication, reflections are particularly bad. As data are coded based on signal pulses and if signal pulses are distorted due to reflections, then data become corrupted. Therefore, reflections should be eliminated.

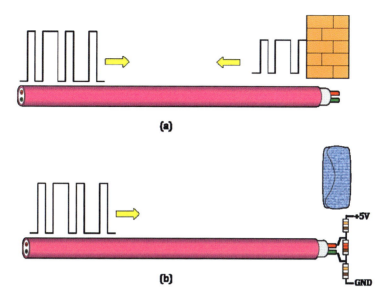

Figure 14 (a): The ends of a cable are of open circuit which reflect signals. (b): One end of the cable is terminated with matching impedance and signals are absorbed.

Any change in resistance, capacitance or inductance causes discontinuity in electrical characteristics and hence causes reflections. In particular the end of a cable is a major area of discontinuity, where the resistance suddenly increases to infinity and the end becomes an open circuit (Figure 14(a)).

To absorb signals at the end of a cable, a circuit of three resistors (Figure 14(b)) is added which provides the "characteristic impedance" that matches the cable impedance. The circuit is called *terminator* and when powered provides the required impedance. As the terminator is powered, the technique is referred to as "active termination". A terminator obtains power either from a dedicated power supply unit or from the PROFIBUS device where the terminator is connected through a sub-D connector.

2.4.1 Termination

A terminator can be either a standalone box (Figure 15) or built into a sub-D connector (Figure 16 (a)). It comprises three resistors, which are 390 Ω, 220 Ω, and 390 Ω respectively.

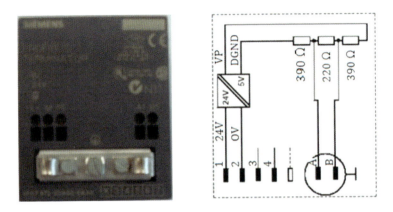

Figure 15: An active terminator

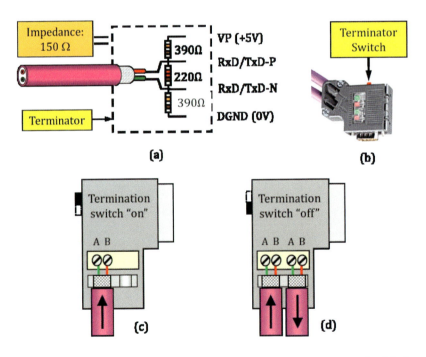

(a) Termination circuit (b) A sub-D connector with a termination switch (c) Connection for the first and last stations in a segment (d) Connections for stations in the middle of a segment

Figure 16: A PROFIBUS DP sub-D connector

The terminator in a sub-D connector obtains power from the device where it is connected and can be switched on or off. When the terminator is on, the bus cable terminates at the connector. When the terminator is off, the cable continues from the entrance of the connector to the exit. This design has a useful feature; that is, when a termination is turned on in the middle of a network (e.g., at Address #10, Figure 17), the downstream nodes (i.e., #11 and #12 of Figure 17), are disconnected from the network. Therefore, by switching a terminator on or off, stations can be removed from or added to the network making troubleshooting and commissioning convenient.

However, the terminator in a sub-D connector is also a potential source of faults because the red switch could be mal-functional. Moreover, when it is used as termination at the end of a segment it will lose power when the last node is removed. It is therefore good practice to use active termination for DP segments.

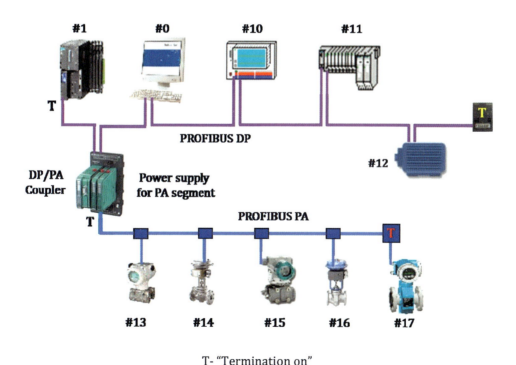

T- "Termination on"

Figure 17: Termination, DP and PA networks

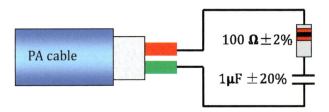

Figure 18: The PA terminator

For PA networks, termination is "passive" or powered from the bus. Figure 18 shows a PA terminator which is a circuit of a resistor and capacitor.

2.4.2 Rules of termination

Two ends of a cable must be terminated and the two terminators must be powered at all times. No terminators should be turned on in the middle of a cable. The rules are simple but they are difficult to obey.

2.5 Spurs

Spurs are also called stubs or drops and they can be used conveniently to connect a PROFIBUS device to a network cable (Figure 19 (a)). However, whether or not a spur can be used depends on the network speed (Table 5). At 1.5 Mbps or lower speeds, spurs can be used but their length is limited. At high speed (higher than 1.5 Mbps), spurs are forbidden and the only connection type permissible is a "daisy chain" using sub-D connectors (Figure 19 (b)).

Allowance of RS485 (DP) spurs is provided in Table 5.

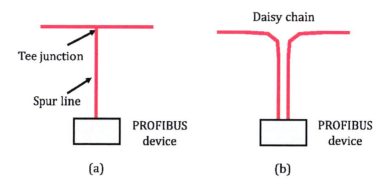

Figure 19: Spur and daisy chain connections

Bit Rate	Total allowable spur capacitance	Total spur cable length per segment
> 1.5 Mbps	None	None
1.5 Mbps	0.2 nF	6.7 m
500 Kbps	0.6 nF	20 m
187.5 Kbps	1.0 nF	33 m
93.75 Kbps	3.0 nF	100 m
19.2 Kbps	15.0 nF	500 m

Table 5: Allowance of RS485 (DP) spurs

Number of spurs	Max. spur length (non-intrinsically safe)	Max. spur length (intrinsically safe)
25 to 32	1 m	1 m
19 to 24	30 m	30 m
15 to 18	60 m	60 m
13 to 14	90 m	60 m
1 to 12	120 m	60 m

Table 6: Allowance of MBP (PA) spurs

From Table 5, it can be seen that 6.7 meters of spurs can be used in a DP segment if the network runs at 1.5 Mbps. For example, if there are 6 motors, then every motor can be connected to the trunk cable via a maximum spur of 1.1 meters. If there are 2 motors, then each spur can be maximum of 3.3 meters.

For PA networks, each spur length is limited as provided in Table 6. For example, if there are 15 transmitters, every transmitter can be connected via a maximum spur of 60 meters giving a total spur length of 900 meters, leaving up to 1000 meters maximum trunk length. The maximum cable length of a PA segment is 1900 meters.

Figure 20 shows two terminators in a PA segment. The first one is inside the coupler, which is always turned on as it is the start of a PA segment. The other terminator can be in the last junction box or at the last station. The last station is the one farthest away among other stations from the last junction box.

If the termination is at the last station, then the cable between the junction box and the last station becomes part of the trunk cable and such an arrangement reduces the number of spurs by one.

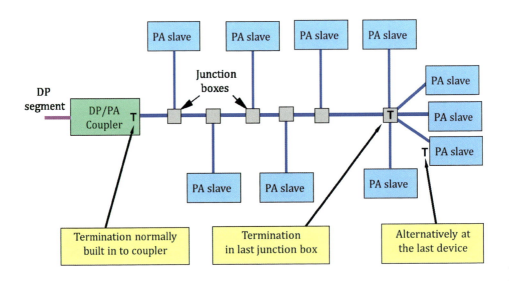

Figure 20: Termination of a PA segment

2.6 Causes of Reflections

Any change in resistance, capacitance or inductance of a PROFIBUS segment will cause reflections. The most common causes are listed below.

2.6.1 Missing terminator

When a long cable is laid, a repeater is duly added and the cable is terminated at the master and repeaters. This is correct and does not normally have many problems in practice. It is also true to say that termination at repeaters is not normally forgotten. However, when OLMs are used, termination at an OLM is often missed! One reason could be that OLMs are often located higher than other PROFIBUS devices and the terminator switch on an OLM is not so easily seen and perhaps often forgotten.

2.6.2 Extra terminator

This can be caused by the terminator in a connector being switched on or the internal terminator in a device that is left on.

2.6.3 Un-powered terminator

A terminator obtains power either from the station that it is connected to or from a dedicated power supply. The former is un-reliable because if the station is removed, the terminator will lose power. The latter is reliable and in some cases redundant power sources are used.

2.6.4 Faulty terminator

A terminator can be damaged or may have deteriorated, especially those in the sub-D connectors.

2.6.5 Spurs

Spurs cause reflections especially at a high speed. How much spur length is allowed in a segment depends on the network speed. Refer to Table 5 and Table 6.

2.6.6 Un-certified device

There is a short cable from the circuit board inside a PROFIBUS device to the device case. This internal cable is a spur and if not certified can cause reflections.

2.6.7 Mixed cables

If different types of cables are used in a segment or in a network, there will be differences in the cable specification. These discrepancies will cause reflections. Therefore, when extending a cable, the new cable must be identical to the existing one in specification.

2.6.8 Short cables

The interface of a PROFIBUS device measures a capacitance of 25 pF approximately and this capacitance must be less than 30 pF according to the PROFIBUS standard (IEC 61158-2). If two devices are connected together with a very short cable as shown in Figure 21, the total capacitance adds up and the reflection also adds up (Figure 21 (b)). This amount of reflection may corrupt communication, e.g., missing or carrying wrong values.

It is recommended by PROFIBUS International that the minimum cable length between any two stations should not be less than 1 meter when network speed is over 1.5 Mbps. This is referred to as the "1-meter rule". Two PROFIBUS stations can be next to each other physically but the connecting cable between them has to be at least 1 meter long. With a longer cable, reflections do not add up but spread apart (Figure 21 (c)).

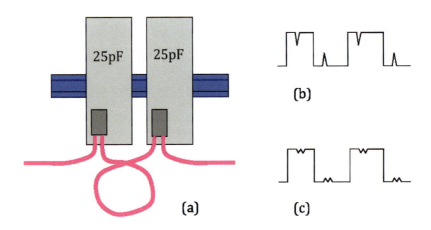

Figure 21: (a) A connecting cable between two stations (b) Reflections add up if the length of a connecting cable is too short (c) Reflections spread apart if the length of a connecting cable is sufficiently long.

3 CONNECTORS AND CABLING TEST

3.1 Connectors

A sub-D connector has 9 pins (Figure 22) . Data pins (Pin 3 and Pin 8) are mandatory (5). Pin 6 and Pin 5 are for fetching power for the terminator from a connected device. Pin 1 should not be used. Instead, the connector case is used for connecting a cable shield. Pins 2 and 7 can be used for supplying 24 DVC to stations on a PROFIBUS DP cable.

Sub-D connectors are rated IP20 and are only suitable for cabinet assembly. For installation in exposed areas, higher ingress protection (Ingress Protection) is required. For example, in the water industry, IP65 and IP67 rated M12 connectors are applied.

Pin No.	Signal	Function	Note
1	Shield	Ground connection	O
2	M24	Ground for +24V supply	O
3	RxD/TxD-P	Data line positive (B-line)	M
4	CNTR-P	Repeater direction control signal	O
5	DGND	Data ground for + 5V	M
6	VP	+5V supply for termination	M
7	P24	+24V supply output voltage	O
8	RxD/TxD-N	Data line negative (A-line)	M
9	CNTR-N	Repeater direction control signal	O
Case	Shield	Ground connection	O

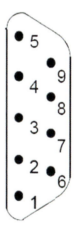

O = Optional   M = Mandatory

Figure 22: Pin allocation, sub-D connectors

M12 connectors for PROFIBUS DP devices have 5 pins as shown in Figure 23 and are referred to as M12 B-coded connectors.

M12 connectors for PROFIBUS PA devices have 4 pins as shown in Figure 24 and are referred to as M12 A-coded connectors. Note that only 2 pins are used.

Solid core cables using "insulation displacement" technology are often used, where the core insulation is not removed, but is pierced through by the blades in a connector. When stripping a cable to fit into a connector, make sure that the cable braids are connected to the connector's screen connection plate as illustrated in Figure 25.

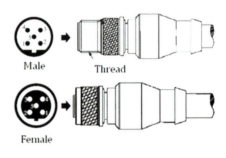

Pin No.	Signal	Function	
1	VP	+5V supply for termination	M
2	RxD/TxD-N	Data line negative (A-line)	M
3	DGND	Data ground for +5V	M
4	RxD/TxD-P	Data line positive (B-line)	M
5	Shield	Ground connection	O
Thread	Shield	Ground connection	O

Figure 23: M12 B-coded male and female connectors for PROFIBUS DP

Pin No.	Signal	Function	
1	RxD/TxD-P	Data line positive (B-line)	M
3	RxD/TxD-N	Data line negative (A-line)	M
4	Shield	Ground connection	O
Thread	Shield	Ground connection	O

Figure 24: M12 A-coded male and female connectors for PROFIBUS PA

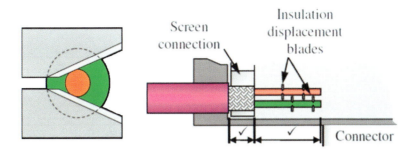

Figure 25: Insulation displacement connections

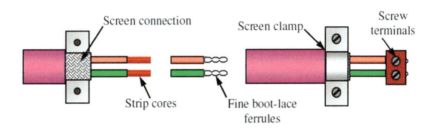

Figure 26: Stranded core connection

Stranded core cables are best used in conjunction with screwed termination connectors (Figure 26). The wire ends are stripped of insulation. Ideally a small (0.75mm) bootlace ferrule should be crimped onto the bare wire using a proper crimp tool (not pliers). The strands should not be twisted together.

Connectors and cables are assembled manually when installing PROFIBUS cables on site. This is where human errors are often introduced. A very common error is that the cable shield is not connected to the shield connecting metal plate as illustrated in Figure 27 (b) and (c), Cable 2. Another common error is that the cable jacket is not clamped at the connector case as shown in Figure 27 (b), Cable 1. Connectors shown in Figure 27 (a) and (d) are correctly assembled.

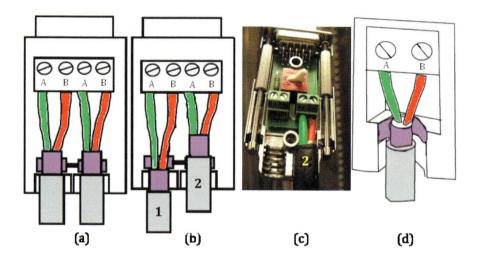

Figure 27: Possible faults inside a connector

Figure 28: A screen clamp for screw terminals

Screen connection into screwed terminals can also pose a difficulty. Do not use a "pig-tail" when connecting screens since this introduces an inductance which creates impedance at high frequencies. A 360° connection of the screen is best. Special stainless steel screen clamps are available which provide a simple and good quality screen connection to screw terminals (Figure 28).

3.2 Cabling Test

Cabling errors can be found using a hand-held cabling tester, e.g., Siemens BT200. Such a tester is used in conjunction with a test plug (Figure 29). The test plug is connected to the first connector and must not be moved around. The terminator of the first connector is switched on. Then, a tester is used to check every connector on a cable starting from the second connector. Where the tester is connected, the connector's terminator is turned on. The terminators of all other connectors are off. Thus, there are only two terminator; one at the first connector and the other where the tester is connected. Only the cable portion within the two terminators is under test and any faults found will be within the two connectors. This is the best way to narrow down faults between two connectors.

A cabling tester can perform the following checks on a PROFIBUS cable:

- Detection of wiring breaks in wire A, wire B or screen
- Detection of short-circuits in or between data lines and screen
- Detection of terminator faults

These tests are typically performed before devices are connected to the network cable and before the cable and devices are powered. They are referred to as "static cabling test" as illustrated in Figure 29.

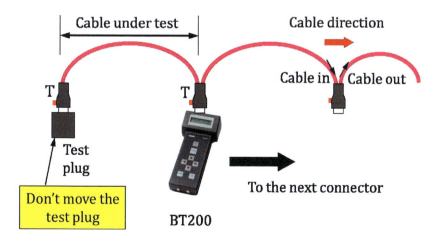

Figure 29: Static cabling test, devices not connected to the cable

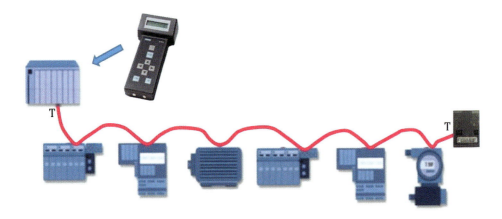

Figure 30: Cabling test, devices connected (2)

When devices are connected, cabling test is performed by replacing the master with a BT200 tester (Figure 30).

3.3 Address Check

A cabling tester can also be used to verify station addresses. This simple verification can detect faulty addressing switches and wrong or duplicated addresses. Addressing verification is performed while the network is powered up. To change a station address, the station has to be powered down and powered up again.

3.4 M12 Connection

In Figure 17, incoming and outgoing cables are joined by a sub-D connector and this allows a PROFIBUS device to be removed or added easily without disconnecting the bus cable. For M12 connection, a bus cable is connected to the devices which typically have two M12 ports (Figure 31). If the two ports are used, the bus cable will be disrupted when the device is removed. This is an issue with M12 connection. The solution is to use just one port via a Tee piece (6). However, a Tee piece introduces a short spur restricting the network speed to 1.5 Mbps or lower. Since M12 devices are mainly used in the water industry or the process industry, 1.5 Mbps or lower is acceptable.

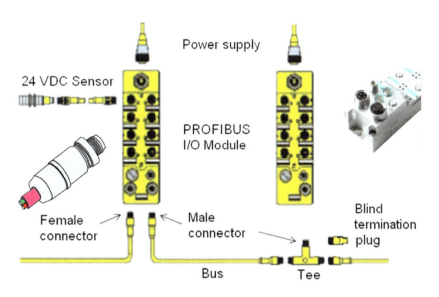

Figure 31: M12 connection

Similarly to sub-D connectors, M12 connectors can also be assembled on site with few special tools required. However, if the routes of cables are known and hence the length of some cables is known, there are pre-assembled M12 cables commercially available. Using pre-assembled cables eliminates many human errors.

4 REPEATERS

4.1 Functions of a Repeater

Figure 32 shows three types of repeaters. A Siemens repeater (a) or Procentec repeater (c) has two built-in termination switches, one for each segment. If there are two cables at one segment as shown in Figure 32 (a), Segment 1, the termination switch of that segment should be off as the cable is continued rather than terminated there. If there is only one cable then the cable end should be terminated, i.e., the termination switch should be on (Figure 32 (a), Segment 2).

Another feature of a Siemens repeater (3) is the built-in piggy back socket that allows a troubleshooting tool to be connected to the top segment of the repeater.

Note that the piggy back socket at a Procentec repeater (Figure 32 (c)) is connected to the bottom segment. Redundant power supplies are used at a Procentec repeater making the repeater more resilient in case one power supply should develop a fault.

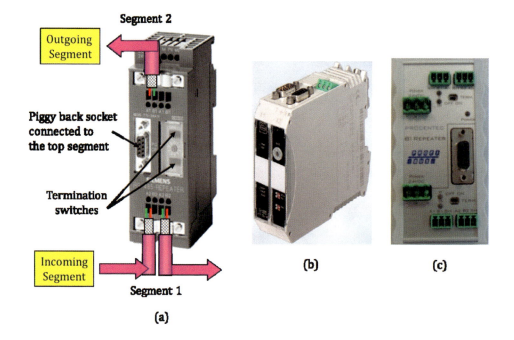

Figure 32: Three types of repeaters

The STAHL repeater (Figure 32 (b)) has no built-in termination switches. It has two RS485 ports for connecting two external sub-D connectors. An external connector will have to have a termination switch and the switch is turned on if there is one cable at the connector and turned off if there are two cables.

A repeater has three functions:

   a. Amplifying signal for long cables
   b. Separating segments
   c. Generating earth free segments (galvanic isolation)

4.2   Examples - Using Repeaters

We look at an example of repeaters being used for segmentation (Figure 33). The engineering station can be easily removed **after** terminating the top segment at Repeater 1. To put the engineering station back into the network, connect the engineering station to Repeater 1 and then switch off the terminator at Repeater 1.

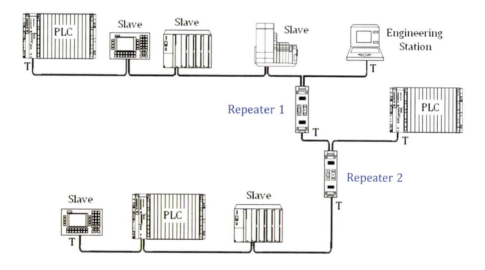

Figure 33: Using repeaters for segmentation

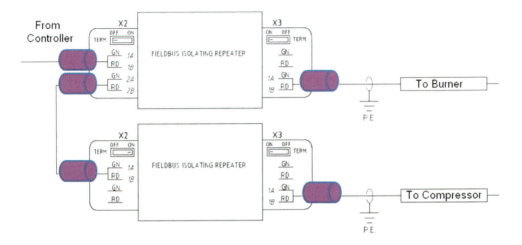

Figure 34: Using repeaters for long cables

Figure 34 shows another example using repeaters for running long cables, e.g., from a control room to a burner or compressor room. Note the position of every termination switch shown in Figure 34.

Galvanic isolation is used where two or more electric circuits must communicate, but their earthing grounds are at different potentials. Figure 35 shows two segments on a network. There is data exchange across the two segments and the segments are electronically separated. Device 1 and the connector, Plug 1, are in Segment 1 and earthed. Device 2, Plug 2, the power supply to the repeater, the rail and Repeater are not earthed or earthed through a capacitor (the RC circuit). This technique is called "capacitive grounding[4]", that uses a repeater and capacitive circuit to separate two earthing systems.

---

[4] Capacitive grounding is also referred to as indirect grounding and more details are included in Section 5.4.

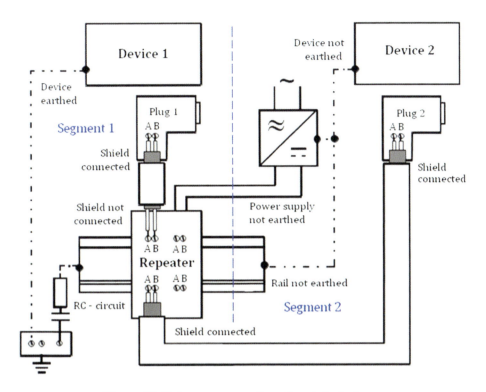

Figure 35: Using a repeater for galvanic isolation

4.3 ProfiHubs

ProfiHubs are products of Procentec (11). Such a hub incorporates several miniature repeaters in a single box. It provides connections for trunk in and trunk out (Figure 36) and several branches, each of which can use the maximum segment cable length and host up to 31 devices. ProfiHub provides very flexible topologies. The five channels or segments are galvanically isolated.

ProfiHubs are grounded in three ways (Figure 37), which are also widely used for other PROFIBUS devices.

(1) direct grounding,
(2) indirect grounding, and
(3) combination of direct and indirect grounding.

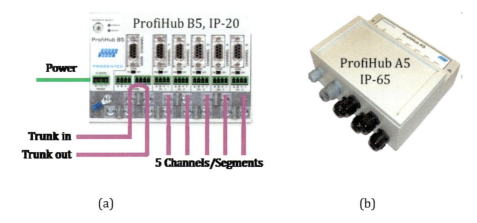

Figure 36: Hubs of repeaters

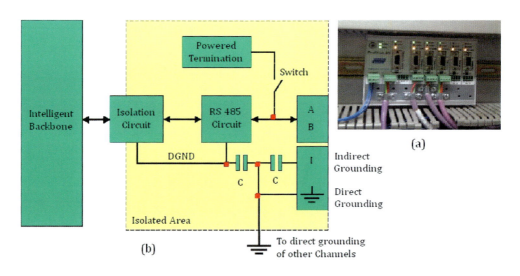

Figure 37: ProfiHub channels and grounding points

It is recommend to always use direct grounding (refer to Section 5.3), which connects the power supply ground point and PROFIBUS cable shield together. If this is not practical, cable shields are connected to indirect grounding points (or capacitive grounding points). If by accident, the direct and indirect grounding points of a ProfiHub are connected together, the direct grounding method wins.

5 ROUTING CABLES AND INSTALLING DEVICES

5.1 Interference

Any industrial environment has electric noises in sources including AC (alternating current) power lines, radio signals and machines, etc. The control of noises in automation systems is vital, as it may become a serious problem even with the best product and technology. Fortunately, simple techniques are available to reduce noise effectively, e.g.:

- adequate grounding, correct shielding methods,
- twisted wires,
- balanced transmission method, and
- filters and differential amplifiers.

In a factory, frequency inverters are widely used and they generate electromagnetic interference (EMI). Their amplifiers typically emit a significant EMI on 10 MHz to 300 HMz frequencies and cause interference in nearby equipment. Complete immunity is not attainable. But, good layout, correct wiring, adequate grounding and shielding techniques offer a significant reduction of such interference.

The measures adopted in PROFIBUS systems to reduce interference are:

- Balanced transmission. PROFIBUS DP and PA both are of balanced transmission methods, which means that the two wires both carry signals. Red wire carries positive signal and green negative.

- The two signal wires are twisted. If the wires are to be exposed, e.g., as a tail for making terminal connections, the tail should be stripped as short as possible so that the exposed and un-twisted wires do not introduce interference.

- The two wires are shielded. If the wires are to be exposed, then the exposed end should be as short as possible. The cable shield should also be grounded according to the grounding scheme specified in design.

- Shielding is a means of weakening EMI. Interference currents on cable shields are diverted to ground via the shield metal bar, which forms a conductive connection with the housing.

5.2 Cables for PROFIBUS DP

The standard, IEC 61158-2, specifies a "Type A" cable for use with PROFIBUS DP as shown in Table 7.

Impedance	135 to 165 Ω at a frequency of 3 to 20 MHz
Capacitance	< 30 pF / m
Resistance	≤ 110 Ω/ km
Wire diameter	> 0.64 mm
Conductor area	> 0.34 mm^2

Table 7: Cable specification for PROFIBUS DP type A cable

In order to fit the cable into standard connectors, the cable needs to have a sheath diameter of 8.0±0.5 mm. The term, "Type A", is rather confusing because it really means "Quality A", i.e., "best quality". Several different forms of Type A cable are available:

- Standard PROFIBUS solid-core cable
- Stranded-core cables for flexibility
- Cables with special sheaths for use in the food and chemical industries
- Armoured cables for protection against rodent and other damage
- Zero Halogen (Low Smoke) cables for use in confined spaces

5.3 Cables for PROFIBUS PA

IEC 61158-2 specifies four different types of cable for use in PA segments (Table 8). PA Type A cable (not the same as DP Type A cable) is a two-core shielded twisted pair cable which gives the best performance in terms of signal attenuation and cable length.

	Type A	Type B	Type C	Type D
Cable description	One pair cable, shielded, twisted	One or more pairs, shielded, twisted	Several pairs, not shielded	Several pairs, not shielded, not twisted
Nominal Conductor Area	0.8 mm² (AWG 18)	0.32 mm² (AWG 22)	0.13 mm² (AWG 26)	0.25 mm² (AWG 16)
DC Resistance Maximum (loop)	44 Ω/km	112 Ω/km	264 Ω/km	40 Ω/km
Characteristic Impedance at 31.25 KHz	100 Ω ± 20%	100 Ω ± 30%		
Maximum Attenuation at 39 KHz	3 dB/km	5 dB/km	8 dB/km	8 dB/km
Maximum Unbalanced Capacitance	2 nF/km	2 nF/km		
Distortion of the Group Delay (7.9 to 39 kHz)	1.7 µs/km			
Area covered by Shield	90%			
Maximum segment cable length (incl. spurs)	1900 m	1200 m	400 m	200 m

Table 8: Cable specification for PROFIBUS PA cables

When using IEC 61158 Type A cable, the total segment cable length can be up to 1900 m including all spurs.

It is recommended to use Type A cable for new PA installations. However, the possibility to other cables is useful when PROFIBUS devices are being added into an existing plant where cables are already installed. The maximum segment cable length significantly reduces when using Type B, C or D cables. Refer to Table 8.

5.4 Installing PROFIBUS Cables and Devices

PROFIBUS controllers, remote IOs, networking components (i.e., repeaters, couplers and OLMs), and drives (or the bus interface modules) are normally installed into cabinets. Motors, pumps, valves, and actuators are normally installed outside those cabinets.

For example as shown in Figure 38, a remote IO station and the drive (not illustrated) for the motor are installed in the cabinet and the motor is in the field outside the cabinet. There could be two scenarios; the motor is either a PROFIBUS device or not a PROFIBUS device. The latter is illustrated in Figure 38. Control signals (turn the motor on or off) and monitoring signals (the motor is on or off) are connected to the remote IO station. If the motor is of PROFIBUS, control and monitoring signals will be transmitted via the bus rather than via the remote IO.

It is typical that a cabinet supplies both high-voltage (380 AC) and low-voltage (24 DC) powers. The power units are earthed to the protective earth (PE), which are circled earth points as shown in Figure 38. Local PE points, e.g., at the Main Rack, Subrack and Motor, should be bonded together using an equipotential bonding cable (5). This provides a base for functional earth (non-circled earth points in Figure 38) and ensures that earth potential at different locations is equalised and there is no current passing through the PROFIBUS cable shield. An earth current clamp meter is required to measure if the earth current is within the specification of 5 ~ 10 mA.

In practice, all PROFIBUS interfaces and cable shields are connected to the functional earth. A grounding cable is laid from a cabinet to another to equalise the earth potential at different locations of the plant or between cabinets within a PROFIBUS segment so that no current flows over the shielding of PROFIBUS cables.

Note that potential equalisation is within a segment of PROFIBUS networks. Segments are formed because:

 a. There are more than 32 stations. Refer to Figure 33.
 b. The network cable is long. Refer to Figure 34 and Table 4.
 c. The need for galvanic isolation. Refer to Figure 35 and Figure 49.

The need for galvanic isolation means that potential equalisation is performed within a segment. A segment can be on its own functional system provided it is galvanically isolated from other local functional systems.

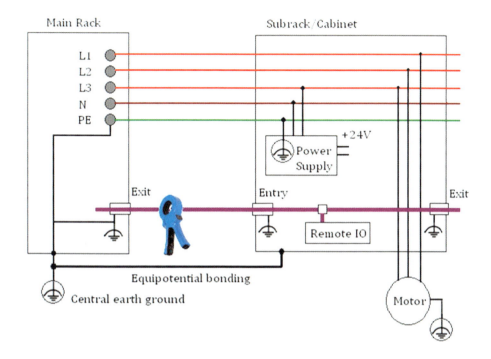

Figure 38: Installing PROFIBUS DP devices

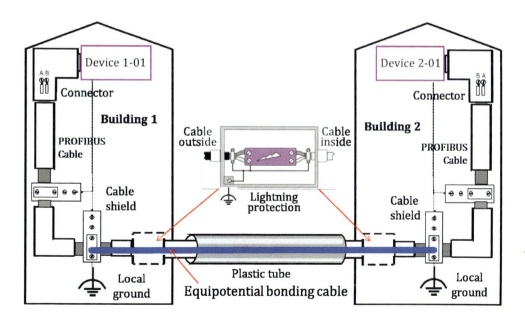

Figure 39: Equipotential bonding between buildings

All PROFIBUS interfaces should be connected to functional earth. The bus cable shield at the exit or entry of a cabinet must be connected to functional earth. Functional earth is eventually connected with protective earth.

Earth potential equalisation is critical to the success of PROFIBUS installations. A bonding cable must satisfy the requirement of IEC 60364-5, which requires that the minimum cross sections of a bonding cable must be:

- Copper, 6 mm²
- Aluminium, 16 mm²
- Steel, 50 mm²

Equipotential bonding should be between cabinets, machines, buildings and sites. Figure 39 shows equipotential bonding between two buildings where devices and cable shields within one building are connected together to the local functional earth.

Cables enter a building through a lightning protection box which separates out-door and in-door cables (Figure 39).

Cable shields are grounded at the exit or entry of a cabinet through strain-relief fittings as shown in Figure 40.

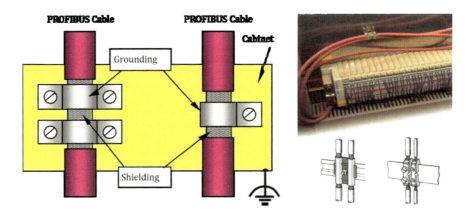

Figure 40: PROFIBUS cables entering or exiting a cabinet

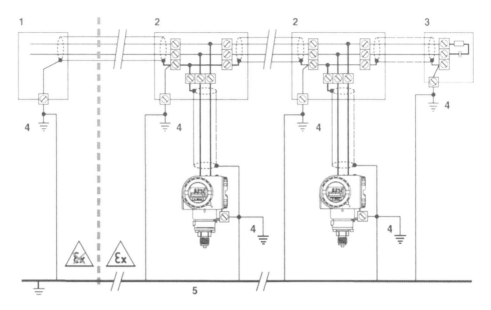

1 – Power Supply or Segment Coupler   2 – Junction Box   3 – Bus Terminator
4 – Local Ground   5 – Equipotential Bonding Cable

Figure 41: Installing PA devices, direct grounding

Generally there is no difference between installing DP and PA systems (9). The shield of a PROFIBUS PA cable is connected to the equipotential bonding system at every device, which is shown in Figure 41. If the bonding cable is in place and there is no difference in earth potentials, the bus cable shield is directly connected to the cases of junction boxes and cases of transmitters. These cases are connected to their local ground and these local grounds are then connected to the equipotential bonding system.

Figure 42 illustrates another grounding scheme – capacitive grounding. The power supply or segment coupler is typically in a cabinet and far away from the site where the junction boxes, transmitters and other PA stations are located. Equal bonding of the earth potentials may not be practical or necessary as, for example, the site is rather wet and there is not much difference in earth potentials. Then, the bus cable shield is connected to a local ground through a capacitor.

"Grounding at one end" means that the bus shield is connected to ground at the power supply as shown in Figure 43 where the ceramic capacitor is ≤ 10nF and test voltage ≥ 1500V.

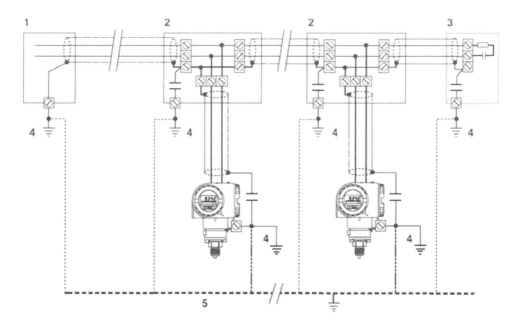

1 – Power Supply or Segment Coupler 2 – Junction Box 3 – Bus Terminator
4 – Local Ground 5 – Equipotential Bonding Cable

Figure 42: Installing PA devices, capacitive grounding

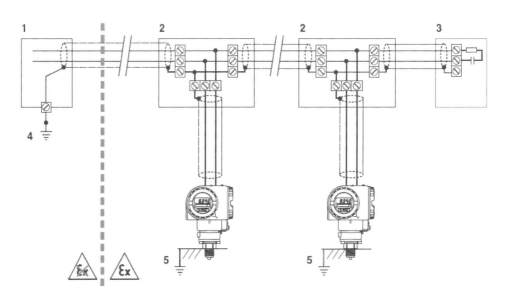

1 – Power Supply or Segment Coupler 2 – Junction Box 3 – Bus Terminator
4 – Grounding point for bus cable shield 5 – Grounding of devices

Figure 43: Installing PA devices, grounding at one end (full isolation)

However, in explosive applications, only direct grounding or grounding at one end are used (Figure 41 and Figure 43).

Note that grounding at one end can only supress low frequency interference. Thus, in a noisy environment, for example, where heavy electromagnetic interference exists, cables are enclosed in a metal conduit or steel tray. A metal tray is normally partitioned and different categories of cables are routed in different partitions. Figure 44 shows a partitioned cable tray and Table 9 lists the cable categories.

Armoured PROFIBUS cables are used if the cable routes are under ground or in exposed areas. Armoured PROFIBUS cables are installed as the same as to other armoured cables.

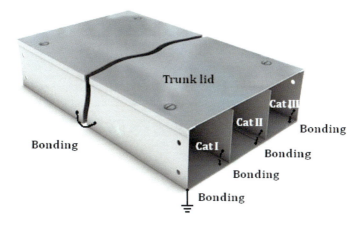

Figure 44: A partitioned cable tray

Figure 45: Armoured PROFIBUS DP cables

Cable Category	Cables
Cat I	Sensitive and safe cables: • Fieldbus and LAN cables (e.g. PROFIBUS, PROFINET AS-i, and Ethernet cables, etc.) • Shielded cables for digital data (e.g. printer, RS232 etc.) • Shielded cables for low voltage ($\leq 25V$) analogue and digital signals even they are in high frequency • Low voltage power supply cables (AC $\leq 25V$ or DC $\leq 60V$). Coaxial signal cables
Cat II	Medium voltage cables: • Cables carrying DC voltages > 60V and ≤ 400 V • Cables carrying AC voltages > 25V and ≤ 400 V
Cat III	High voltage interference sources: • Cables carrying DC or AC voltages > 400 V • Cables with heavy currents Motor/drive/inverter cables • Telephone cables (can be transiently >2000V)
Cat IV	• Any cables at risk of direct lightning strikes, e.g., cables running outside, between buildings

Table 9: Cable categories

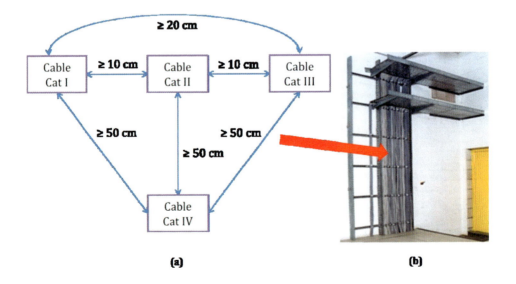

Figure 46: Cable separation

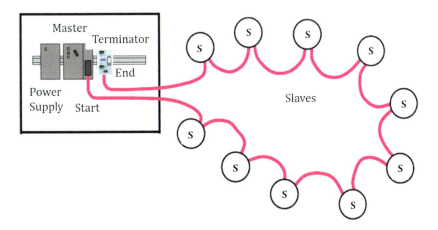

Figure 47: The start and end of a bus cable

If there is a single tray (Figure 46 (b)), the different categories should be separated by the recommended distances (Figure 46 (a)).

For maintenance purposes, if possible, it is recommended to bring the end of a bus cable back to its start (Figure 47). This makes it easy, for example, to check if power supply to the terminator or the terminator switch are in working condition.

6 OPTICAL TRANSMISSION

It is recommended to use optical fibres between buildings and between different sites (Figure 49). Within a building, fibres are then changed to copper wires using optical link modules (OLMs). Between two OLMs, there will be a fibre cable for transmitting data and the other for receiving data. A copper segment typically ends or begins at an OLM where the termination switch must be turned on, which is illustrated in a circled T in Figure 48.

An OLM is a repeater with both copper and fibre interfaces, which refreshes the signals received on one port and then repeats them on all other ports.

Plastic fibres are often of low cost, simple to make up, but generally limited to distance of less than 100 meters. Multi-mode glass fibres (which allow multiple paths for light to be transmitted) can be used for distances up to 3 km while single-mode glass fibres up to 50km. Refer to Table 10 (7).

As optical fibre cables are completely insensitive to electromagnetic interference (EMI) and also emit no electromagnetic waves into their environment, they are an excellent alternative to copper cables in heavy electromagnetic environment and to save on complex grounding protection schemes including equipotential bonding.

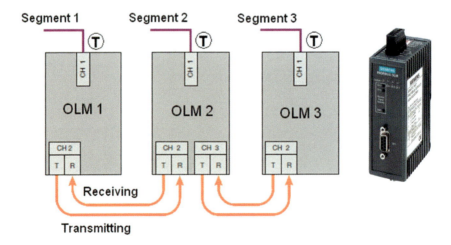

Figure 48: Optical fibre

Fibre Type	Core / Coating Diameter (µm)	Transmission Range
Plastic Fibre	980 /1000	max. 100 m
HCS® Fibre	200 / 230	max. 500 m
Multi-mode Glass Fibre	62.5 / 125	2 – 3 km
Single-mode Glass Fibre	9 / 125	> 15 km

Table 10: PROFIBUS optical transmission

In Figure 49, Segment 1 and Segment 2 are in different sites and they are connected via an optical fibre cable. The two sites are not earth-potential equalised, meaning they have their own local earthing systems. Segment 2, which comprises Device Y, a connecting cable, OLM 2, and the rail, is earthed via a capacitor (1µF) and a resistor (100 Ω). Hence, the two segments are galvanically separated through indirect grounding.

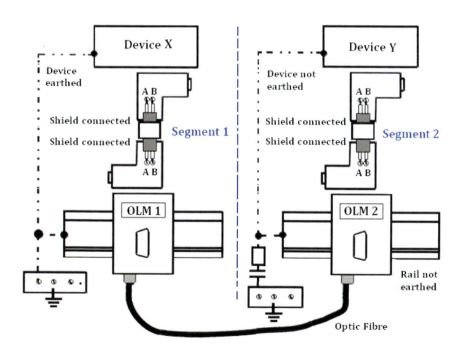

Figure 49: OLMs for galvanic separation

7 SPOT THE ERRORS

This section contains images of some PROFIBUS systems and there are errors. Can you spot the errors? The answers are provided online in the Certified PROFIBUS Installer Course.

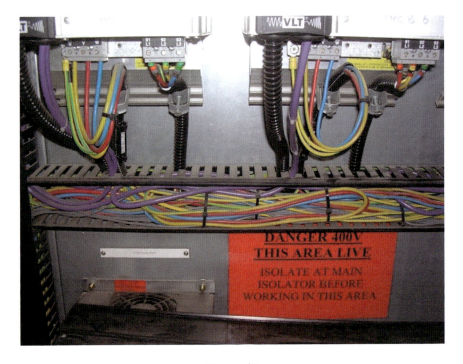

Figure 50

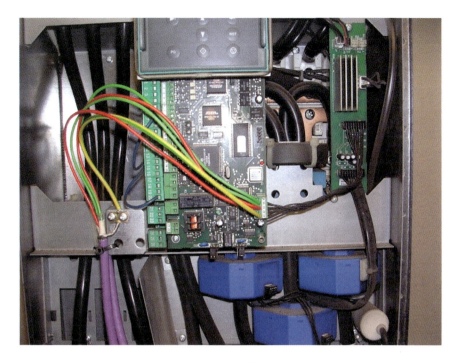

Figure 51

Figure 52

Figure 53

Figure 54

Figure 55

Figure 56

Figure 57

Figure 58

Figure 59

Figure 60

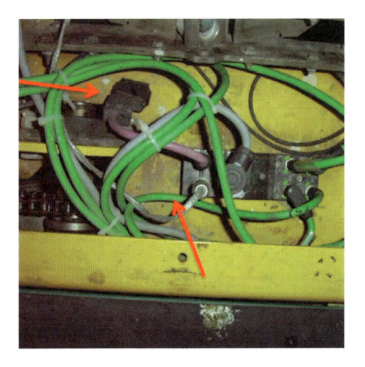

Figure 61

Figure 62

Figure 63

Figure 64

Figure 65

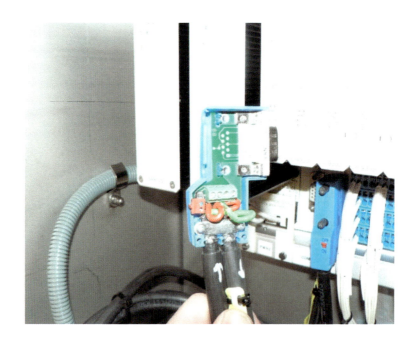

Figure 66

Figure 67

Figure 68

Figure 69

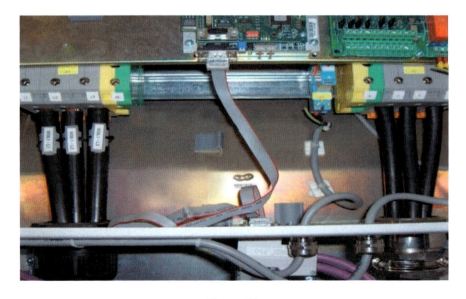

Figure 70

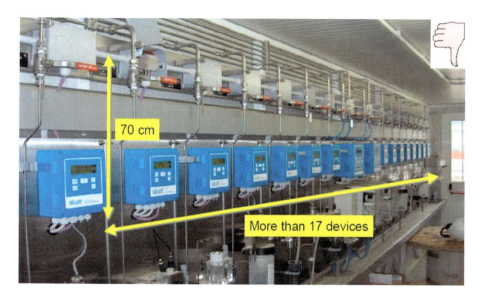

Figure 71

Figure 72

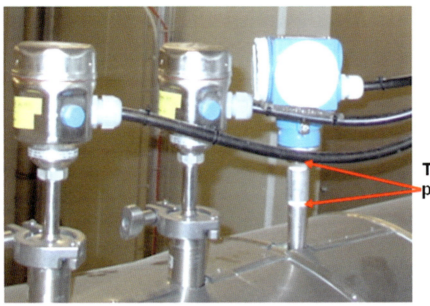

Figure 73

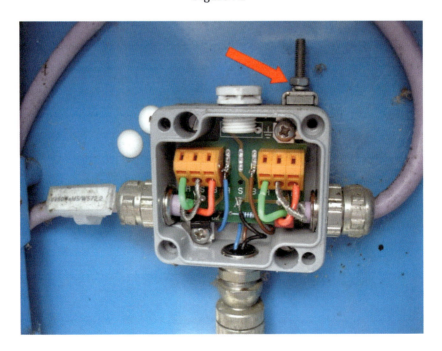

Figure 74

8 INSTALLATION ACCEPTANCE

PROFIBUS is a mature technology and none of the typical main failure modes result from a deep technical problem. Often the real problem is simply a bulge connector or rogue terminator. DIP switches can be as unreliable as terminator switches so false addresses can exist even though the slave settings look correct. However finding the root of the problem on a real network can be exceptionally difficult and time-consuming. Therefore, the very first step in installation acceptance testing is to perform visual inspection on wiring, earthing, segmentation and bus termination.

These visual checks are so important and effective that they are used as part of contractual test documents. This section provides some sample check lists that can be used as a template to build up further installation acceptance documents.

8.1 Network Drawing and Topology

The checks listed here are to obtain the network topology and to produce a network drawing that is required for further tests. If the network drawing is available, it is still required to go through these tests (Table 11) and to ensure that all details are included on the drawing.

System name:			Network name:
Location:			Transmission speed:
Installation acceptance performed by:			Date:
Comments:			
No.	**Answer**	**Test**	
1.	$M_n =$	How many masters are in this network?	
2.	$RPT_n =$	How many repeaters (include diagnostic repeaters and/or Profihubs) are in this network?	

3.	$SG_{n\text{-}DP} =$	Based on the above questions, the number of the DP segments is known.
4.	$SG_{n\text{-}PA} =$	How many PA segments are in this network? (A PA segment starts from a DP/PA Coupler or Link Module.)
5.		How many slaves are in every segment? $S_{n\text{-}DP1} =$ $S_{n\text{-}DP2} =$ … $S_{n\text{-}PA1} =$ $S_{n\text{-}PA2} =$
6.		For every DP segment, measure the cable distance between every two stations and add them together to obtain the: Total cable length in DP Segment 1 = Total cable length in DP Segment 2 = …
7.		For every PA segment, measure the cable distance between any two junction boxes and any drop cable length. Add the trunk cable and drop cable length together. Total cable length in PA Segment 1 = Total cable length in PA Segment 2 = …
8.	Yes/No Yes/No Yes/No	Draw the network layout based on the above information, mark stations with their addresses and indicate masters, slaves, repeaters, DP/PA couplers, and terminators using different symbols. The network drawing is now available. The DP cable length between any two connectors/stations is marked on the drawing. The trunk or drop cable length is marked on the drawing.

Table 11: Network topology and drawing checklist

8.2 Checklist for PROFIBUS DP (RS485) Grounding

These checks are to ensure that grounding of the PROFIBUS stations and bus shield is adequate.

System name:			Network name:
Segment name and location:			Transmission speed:
Installation acceptance performed by:			Date:
Comments:			
No.	**Note**	**Test and Acceptance**	
1.	Yes/No	Is an earth potential equalisation bonding cable installed?	
2.	Yes/No	Is the functional earth in place?	
3.	Yes/No	Are all PROFIBUS station interfaces connected to the functional earth?	
4.	Yes/No	Is the bus shield connected to the functional earth at every station?	
5.	Yes/No	Is the bus shield connected to the functional earth at every entry and/or exit of the cabinets?	

Table 12: Checklist for PROFIBUS DP (RS485) grounding

8.3 Checklist for PROFIBUS PA (MBP) Grounding

These checks are to ensure that grounding of PROFIBUS PA stations and bus shield is adequate.

System name:		Network name:	
Location:		DP transmission speed:	
PA segment name:		Coupler Voltage Rating:	
		Coupler Current Rating:	
Installation acceptance performed by:		Date:	
Comments:			

No.	Note	Test and Acceptance
1.	Yes/No	Is the direct grounding method adopted?
2.	Yes/No	Is the capacitive grounding method adopted?
3.	Yes/No	Is the full isolation grounding method adopted?
4.	Yes/No	Is the DP/PA Coupler grounded according to the grounding method?
5.	Yes/No	Are the PA stations grounded according to the grounding method?
6.	Yes/No	Are the junction boxes grounded according to the grounding method?
7.	Yes/No	Is the bus shield grounded according to the grounding method?

Table 13: Checklist for PROFIBUS PA (MBP) grounding

8.4 Checklist for PROFIBUS DP (RS485) Cabling

Please make sure that the DP cable is not connected to any station and repeat the following tests for every segment.

Segment name:		Location:	
Installation acceptance performed by:		Date:	
Comments:			

No.	Note	Test and Acceptance
1.	Yes/No	Does the cable satisfy the specification of a PROFIBUS DP Type A cable?
2.	Yes/No	Check the direction of the cable. The cable should go into a sub-D connector, go out from the connector and then go into the next connector. Is this direction consistent at every connector?
3.	Yes/No	Are any of the connectors too small or too large to fit into the space, station or machine?
4.	Yes/No	Is the cable length between any two connectors longer than 1 meter?
5.	Yes/No	What is the cable length between any two connectors? Does this length match the distance of the two stations which are to be connected? Total cable length of this segment =
6.	Yes/No	Is the total cable length within the maximum segment cable length limit?
7.	Yes/No	Does the first connector have a piggy back?

8.	Yes/No Yes/No Yes/No Yes/No Yes/No Yes/No Yes/No Yes/No Yes/No Yes/No Yes/No	Using a handheld tester (e.g. Siemens BT200) to check every connector if there are following faults. The two wires are twisted. Has twisting damaged any of the two cores? Is the A line or B line bent sharply in a connector? Is the A line or B line squashed in a connector? Are A and B swapped? Is the A line or B line broken? Is the A line or B line shorted? Is the bus shield broken? Is the A line or B line shorted with the shield? Is the bus shield shorted? Is the connector bulged? Is every connector of the cable checked?
9.	Yes/No	Are all terminators on the cable switched off?
10.		Is the last leg of the cable connected to a repeater or an active terminator?
11.	Yes/No	Confirm that the terminator at the first and last end of the cable is switched on.
12.	Yes/No	Confirm that the two terminators are marked on the network drawing.

Table 14: Checklist for PROFIBUS DP (RS485) cabling

8.5 Checklist for PROFIBUS PA (MBP) Cabling

For PROFIBUS PA cabling test, refer to the section, Testing the PROFIBUS PA cable and the bus connectors, Page 89 of (4).

9 REFERENCES

1. **Powell, James and Vandelinde, Henry.** *Catching the Process Fieldbus : An Introduction to Profibus for Process Automation.* s.l. : Momentum Press, 2012. ISBN-10: 1606503960, ISBN-13:978-1606503966.

2. **Felser, Max.** *PROFIBUS Manual.* Berlin : epubli GmbH, 2011. ISBN 978-3-8442-1435-2.

3. **Siemens AG.** *SIMATIC NET PROFIBUS Networks Manual.* Nuremberg : s.n., 2009. C79000-G8976-C124-03.

4. **PROFIBUS International.** *Installation Guideline for Commissioning.* Karlsruhe : PROFIBUS International, 2006.

5. **PROFIBUS International.** *Installation Guideline for Cabling and Assembly.* Karlsruhe : PROFIBUS International, 2006.

6. **PROFIBUS International.** *PROFIBUS Interconnection Technology Guideline.* Karlsruhe : PROFIBUS International, 2007.

7. **PROFIBUS International.** *PROFIBUS Basic Slide Set.* Karlsruhe : PROFIBUS International, 2010 - 2013.

8. **PROFIBUS International.** *Installation Guideline for PROFIBUS-DP/FMS.* Karlsruhe : PROFIBUS International, 1998.

9. **PROFIBUS International.** *PROFIBUS PA User and Installation Guideline.* Karlsruhe : PROFIBUS International, 2003.

10. **Popp, Manfred.** *The New Rapid Way to PROFIBUS DP (from DP-V0 to DP-V2).* Karlsruhe : PROFIBUS International, 2003.

11. **Procentec.** *ProfiHub A5/B5 User Manual.* Wateringen : Procentec, 2012.

12. **Mitchell, Ronald W.** *PROFIBUS: A Pocket Guide.* s.l. : ISA, 2004. ISBN-13:978-1556178627.

10 INDEX

"Type A", 40
1-meter rule, 26
active termination, 19
address, 7, 8, 16
Allowance of RS485 (DP) spurs, 23
BT200, 31
bus cycle time, 10
Cable categories, 48
cable specification, 40, 41
Cable specification for PROFIBUS PA, 41
Cabling test, 32
capacitive grounding, 36, 37, 38, 45
Class 1, 5
Class 2, 5
Coupler, 4, 16, 17, 45, 46
cyclic, 10
device identification number, 6
DIP switches, 7
direct grounding, 37, 38
DP termination circuit, 20
EExi, 17
EMI, 39
equipotential, 42, 45, 50
explosive, 17, 18, 47
factory automation, 1
field devices, 1
FISCO, 18
functional earth, 42, 44
galvanic isolation, 35
GSD, 6, 73
IEC 60364-5, 44
IEC 61158, 41
IEC 61158-2, 18, 26, 40
indirect grounding, 37, 38
insulation displacement, 28
intelligent device, 1
intrinsically safe, 17
Link Module, 4, 16
M12 A-coded, 28
M12 B-coded, 28
M12 connection, 32
M12 connectors, 28
Master, 5
maximum cable length of a PA segment, 24
MBP, 3, 4, 14, 23
metal conduit, 47
minimum cable length, 26
multi-mode glass fibre, 50
OLM, 14, 16, 25, 50, 51
Pin allocation, 27
plastic fibre, 50
process automation, 1
PROFIBUS network bit rates, 9
protective earth, 42
Reflections, 13, 18
remote IO, 3
repeater, 14, 16, 34, 35, 36, 37, 50
RS485, 3, 4, 14, 23, 35
RS485 load, 14
Screen clamp, 30
segment, 14
Set Slave Address, 8
single-mode glass fibre, 50
Slaves, 5
special DP/PA convertor, 16
Special PROFIBUS addresses, 8
special repeaters, 16
spurs, 22, 23
start-up procedure, 10
Start-up procedure, 10
Static cabling test, 31
steel tray, 47
stub-line, 41
sub-D connector, 13, 19, 21, 27, 32
terminator, 19, 20, 21, 25, 31, 49
token passing, 11

11 GLOSSARY

AC	Alternating Current
AWG	American Wire Gauge
DC	Direct Current
DCS	Distributed Control System
DIP	Dual In-line Package
DP	Decentralised Periphery
EExi	European Standard for Explosive Applications Intrinsic Protection Technique
EMI	Electromagnetic Interference
FA	Factory Automation
GSD	General Station Description
HMI	Human Machine Interface
IO	Input and Output
IS	Intrinsic Safe
IP	Ingress Protection
MCC	Motor Control Centre
MBP	Manchester Bus Powered
OLM	Optical Link Module
PA	Process Automation
PE	Protective Earth
PLC	Programmable Logic Controller
PROFIBUS DP	PROFIBUS Decentralised Periphery
PROFIBUS PA	PROFIBUS Process Automation
PSU	Power Supply Unit
RS485	Recommended Standard 485
SCADA	Supervisory Control and Data Acquisition
SSA	Set Slave Address

ABOUT THE AUTHOR

Dr. Xiu Ji is a Senior Lecturer teaching industrial networks at the Manchester Metropolitan University. Prior to the academic post, she spent many years in the automation industry as a principal system engineer and project manager.

Made in the USA
Lexington, KY
22 August 2014